From Your Friends at The Mailbox®

Math
MIND BUILDERS
Grade 1

Welcome to *Math Mind Builders*! This must-have resource is sure to reinforce math skills while developing critical-thinking skills. Packed with curriculum-based problems and puzzles covering a variety of math topics, this resource provides students with a school year's worth of problem-solving opportunities.

Project Manager:
Njeri Jones Legrand

Editor:
Susan Walker

Writer:
Stephanie Turner

Art Coordinator:
Pam Crane

Artist:
Nolan Galloway

Cover Artists:
Nick Greenwood, Clevell Harris, Kimberly Richard

www.themailbox.com

©2001 by THE EDUCATION CENTER, INC.
All rights reserved.
ISBN #1-56234-418-8

Except as provided for herein, no part of this publication may be reproduced or transmitted in any form or by any means, electronic or mechanical, including photocopying, recording, or storing in any information storage and retrieval system or electronic online bulletin board, without prior written permission from The Education Center, Inc. Permission is given to the original purchaser to reproduce patterns and reproducibles for individual classroom use only and not for resale or distribution. Reproduction for an entire school or school system is prohibited. Please direct written inquiries to The Education Center, Inc., P.O. Box 9753, Greensboro, NC 27429-0753. The Education Center®, *The Mailbox*®, and the mailbox/post/grass logo are registered trademarks of The Education Center, Inc. All other brand or product names are trademarks or registered trademarks of their respective companies.

Manufactured in the United States
10 9 8 7 6 5 4 3 2

INCLUDED IN THIS BOOK

Each activity page features five mind-building math problems plus a more difficult bonus builder problem to boost students' critical-thinking skills. Inside you will find an assortment of problems designed to reinforce the math topics and skills that you teach. Featured topics and skills include the following:

- computation
- numeration
- geometry
- time
- money
- fractions
- place value
- patterns & relationships
- measurement
- problem solving
- graphing, probability, & statistics

HOW TO USE THIS BOOK

Use the activity pages in this book in a variety of ways to supplement your math curriculum.

 For independent practice, duplicate the activity pages for students to use as morning work, problems of the day, free-time activities, or daily homework practice.

 For partner or small-group practice, duplicate a desired activity page and give each pair or group a copy. Have students discuss possible strategies for solving the problems.

 For whole-group problem-solving practice, make transparencies of the activity pages.

 For a learning center activity, duplicate, laminate, and cut apart the activity pages. Group the resulting cards by topic and place specific skill cards at a center. Or, for a mixed review, place a variety of skill cards at a center.

 For assessing students' understanding of math concepts, make individual student copies and have each student explain in writing how to solve each problem.

Numeration

Count the circles. Draw a group of squares with 1 more.

1

Patterns & Relationships

Finish the pattern.

2

Computation

How many birds will be left if 2 fly away?

3

Geometry

How many circles are there?

4

Measurement

Who is shorter: you or your teacher?

5

Bonus Builder #1

How many shells will there be if you draw 1 more?

Computation

COMPUTATION

Count the cookies. If you get 3 more, how many will you have?

(6)

MONEY

Add 3¢. How much money is there now?

(7)

TIME

How many days are in 1 week?

(8)

PROBLEM SOLVING

Which has more legs? How many more legs does it have?

(9)

PATTERNS & RELATIONSHIPS

Complete this pattern.

7, 6, 5, ___, 3, 2, 1

(10)

BONUS BUILDER #2

Look at the graph. Write 2 things you learn from it.

Sara's Bears 🐻🐻🐻
Tom's Bears 🐻🐻🐻🐻

GRAPHING, PROBABILITY, & STATISTICS

Time

What 2 times of the day could this be?

11

Numeration

Put these numbers in order.

7, 6, 8

12

Patterns & Relationships

Complete this pattern.

Friday, Monday, Thursday,

_____

13

Computation

I am 8 more than 3. What number am I?

14

Problem Solving

Make equal sets. How many flowers will be in each vase?

15

Bonus Builder #3

How many paper clips wide are 2 books?

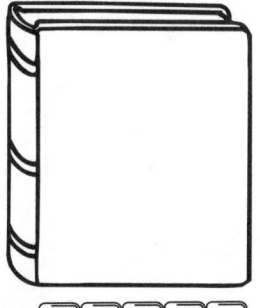

Measurement

Numeration

How many pairs of eyes are there?

16

Money

Circle the coin that is worth more than 5¢.

17

Graphing, Probability, & Statistics

Which takes the least amount of time to do?

18

Geometry

I have 3 sides. Draw my shape.

19

Time

What time will it be in 2 hours?

20

Bonus Builder #4

Take away 1 ten. What number is left?

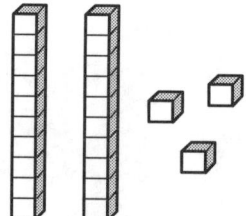

Place Value

MONEY

What 2 coins equal 6¢?

(21)

MEASUREMENT

How many blocks longer is the pencil?

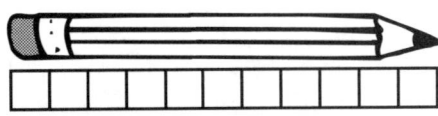

(22)

COMPUTATION

Write the answer.

2 + 1 + 1 − 2 = _____

(23)

GEOMETRY

I am a flat, round shape with no straight sides or corners. What am I?

(24)

PLACE VALUE

If you add 7 ones, what number will you have?

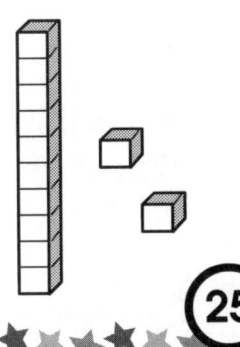

(25)

Bonus Builder #5

What time was it 1 hour ago?

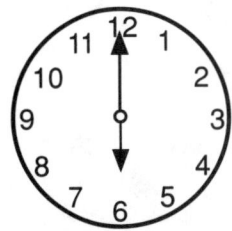

TIME

COMPUTATION

How many more shoes than socks are there?

(26)

GEOMETRY

Draw a shape that has 2 long sides and 2 short sides.

(27)

TIME

How many days are there in 2 weeks?

(28)

PATTERNS & RELATIONSHIPS

Complete this pattern.

(29)

NUMERATION

Which groups are not in pairs?

(30)

Bonus Builder #6

With your eyes closed, which color jelly bean would you probably grab most often? Why?

GRAPHING, PROBABILITY, & STATISTICS

Place Value

Add 2 tens and 4 ones. What is the new number?

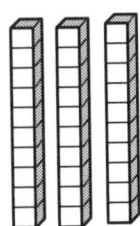

31

Graphing, Probability, & Statistics

Who spent only 2 hours on chores?

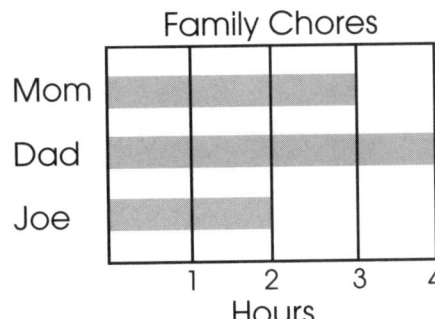

32

Money

Take away 7¢. How much is left?

33

Measurement

What is the total length of the ants?

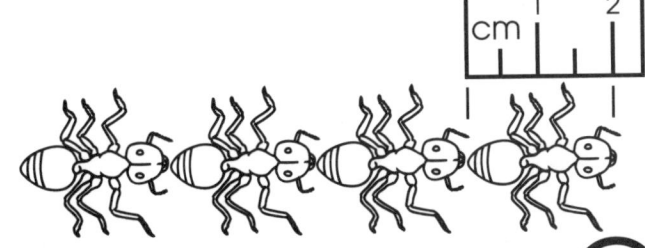

34

Fractions

How many slices are there in half the pizza?

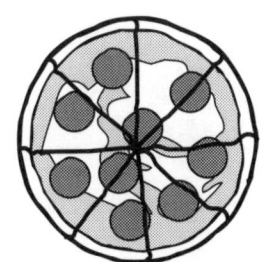

35

Bonus Builder #7

Each basket holds 5 apples. How many more baskets are needed?

Problem Solving

Patterns & Relationships

Finish this pattern.

36

Graphing, Probability, & Statistics

If each ○ = 3 wins, how many games did the Cubs win in all?

Bears   Cubs   Lions

37

Problem Solving

There are 16 kids at the zoo. Seven of them are boys. How many are girls?

38

Money

Draw 3 ways to make 10¢ without using a dime.

39

Computation

Count the marbles. If you lose 2 and buy 3 more, how many will you have?

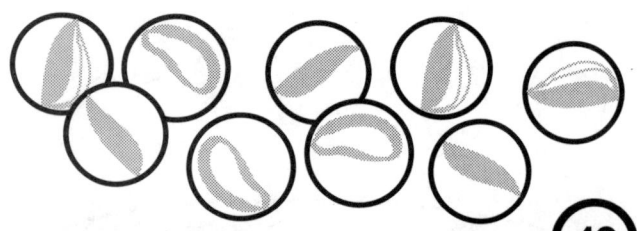

40

Bonus Builder #8

Write the number for each word.

ten
twelve
eight
sixteen

Numeration

Patterns & Relationships

Complete the pattern.

🐜🐜🦋🦋🐜🦋🦋🐜🐜🦋🦋🐜🦋🦋🐜🐜🦋🦋 __ __

41

Measurement

What would you use to measure the *length* of a book?

42

Geometry

Draw a picture using only these shapes.

43

Money

If Ann collects 2 pennies every day for 8 days, how much money will she have?

44

Fractions

Draw a house with 8 windows. Color 1/2 of the windows yellow.

45

Bonus Builder #9

Look at the time. If it takes you 1 hour to dress, eat, and ride to school, will you make it by 8:00?

Time

©2001 The Education Center, Inc. • Mind Builders • Math • TEC1600 • Key p. 45

NUMERATION

Which is the smaller amount?

32¢   23¢

(46)

GRAPHING, PROBABILITY, & STATISTICS

Which 2 kids got the same number of bugs the second week?

Bugs Collected

Names	Week 1	Week 2	Week 3
Tina	5	7	3
Jo	6	7	4
Mindy	3	5	3

(47)

GEOMETRY

Draw a picture of the heart folded in half.

(48)

PLACE VALUE

Group these eggs into tens and ones.

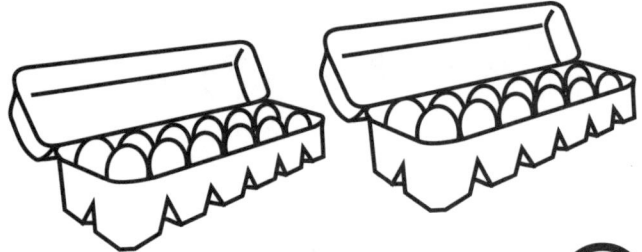

(49)

FRACTIONS

Draw a picture of 3 birds. Color 1/3 of the birds blue.

(50)

BONUS BUILDER #10

You buy 2 bananas and 1 orange. What is the total cost?

COMPUTATION

Problem Solving

How many more pennies do you need to make 3 groups of ten?

51

Time

If today is Wednesday, what day of the week will it be in 6 days?

52

Money

Add 2 nickels to the group of pennies. How much money is there in all?

53

Graphing, Probability, & Statistics

Look at the game. Which number will you probably land on the least?

54

Measurement

Draw a caterpillar that is 5 centimeters long.

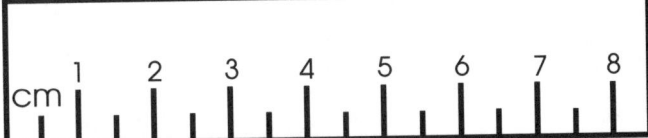

55

Bonus Builder #11

Complete this pattern.

Patterns & Relationships

COMPUTATION

How much money do you need to buy them all?

56

GRAPHING, PROBABILITY, & STATISTICS

Which shape will you land on most often? Why?

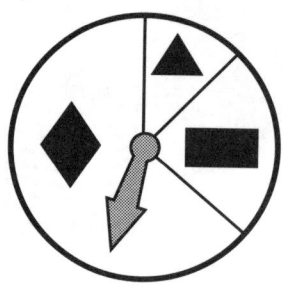

57

PATTERNS & RELATIONSHIPS

Complete this pattern.

1, 3, ___, 7, 9, ___

58

NUMERATION

What number is larger than 20 and between 19 and 22?

59

PLACE VALUE

How many tens are there in 60?

60

BONUS BUILDER #12

Draw 4 erasers. Color 1/2 of them orange. Color 1/4 of them black. What fraction is left uncolored?

FRACTIONS

Time

What time will it be in 4 hours?

(61)

Numeration

Draw a picture in which the girls outnumber the boys by 1.

(62)

Measurement

What do you measure using a scale?

(63)

Computation

Write the answer.

$6 + 2 + 1 - 6 =$ _____

(64)

Problem Solving

Each child needs 5 cards to play. How many more cards are needed?

(65)

Bonus Builder #13

Draw a graph showing that you are asleep for 10 hours and you are awake for 14 hours.

Graphing, Probability, & Statistics

Numeration

Jane won first place in a race. What place did Tony get?

66

Money

How much is 3¢ less?

67

Graphing, Probability, & Statistics

What 3 things can you learn from this graph?

Books Read

	Monday	Tuesday	Wednesday
1st-grade class	20	24	25
2nd-grade class	16	25	18

68

Geometry

How many rectangles do you see?

69

Fractions

How many halves are there?

70

Bonus Builder #14

Count the cookies. If you and 2 friends each ate 2, how many cookies would be left?

Problem Solving

Patterns & Relationships

What comes next? Why?

 _____

71

Measurement

Which is heavier?

 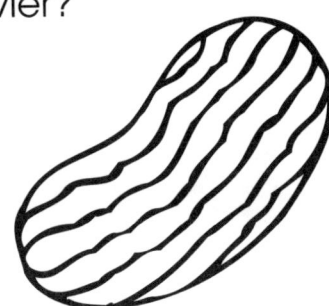

72

Numeration

How many kids are hiding?

73

Money

Draw 4 coins that add up to 20¢.

74

Place Value

Which number is in the tens place?

25

75

Bonus Builder #15

What time was it a half hour earlier?

Time

Geometry

List 3 shapes that have 4 sides.

76

Place Value

Trade the numbers in the tens and ones columns. What is the new number?

125

77

Computation

If you earn 5¢ on Wednesday and 9¢ on Friday, will you have enough money to buy this pencil?

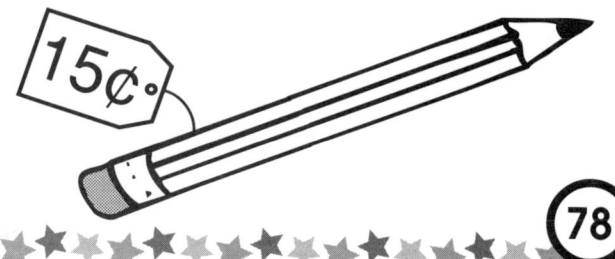

78

Fractions

What fraction of the pizza has been eaten? Show 2 ways to write it.

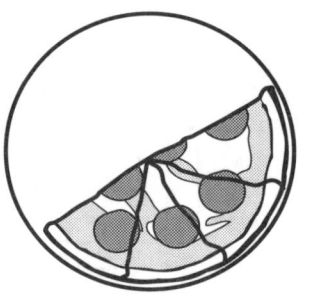

79

Numeration

List all the odd numbers between 0 and 10.

80

Bonus Builder #16

If you don't look inside the bags, from which bag will you be more likely to grab a dark gumball?

Graphing, Probability, & Statistics

Graphing, Probability, & Statistics

How many more blocks did Jan use than Curtis?

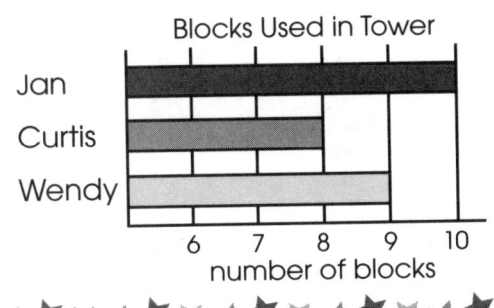

81

Time

What time will it be in 30 minutes?

82

Money

How many nickels equal this amount?

83

Measurement

Which item weighs more?

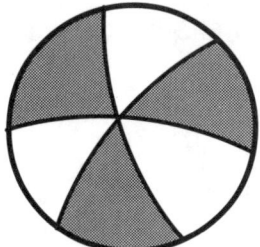

84

Patterns & Relationships

Complete this pattern.

12, 10, ____, 6

85

Bonus Builder #17

I am an odd number between 6 and 16. I have a 3 in the ones place. What number am I?

Numeration

Place Value

Which number does not belong? Explain why.

14   18   26   16

(86)

Fractions

If you and 2 friends each eat a slice, what fraction of the apple will remain?

(87)

Problem Solving

Each child needs the same number of peanuts. How many should each child have?

(88)

Graphing, Probability, & Statistics

If you reach in the bowl without looking, will you probably grab a square or round piece of candy?

(89)

Computation

If you add me to 5, the sum is 13. What number am I?

(90)

Bonus Builder #18

How many sides do a triangle and rectangle have in all?

Geometry

Patterns & Relationships

Complete this pattern.

1, 2, 2, 3, 3, 3, 4, 4, ___, ___

91

Measurement

List 3 things in the classroom that are wider than 3 inches.

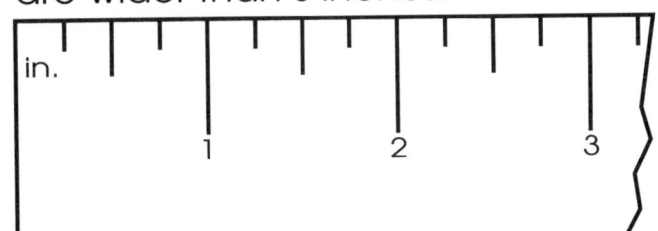

92

Geometry

Find and list 3 spheres in your classroom.

93

Time

How many minutes are there from 1:00 P.M. to 2:00 P.M.?

94

Computation

The kids have 11 pets in all. Carmen has 4 kittens. Sue has 2 dogs. How many fish does Lee have?

95

Bonus Builder #19

I am 2¢ less than the sum of 2 dimes. How much money am I?

Money

©2001 The Education Center, Inc. • Mind Builders • Math • TEC1600 • Key p. 46

NUMERATION

List all the even numbers between 21 and 33.

(96)

GRAPHING, PROBABILITY, & STATISTICS

Look at the graph. Who spends more time outside: Sue and Cal or Tammy and Otis?

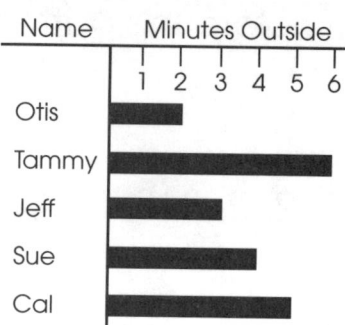

(97)

PROBLEM SOLVING

Each mother cat has 4 kittens. How many mother cats are there for these kittens?

(98)

FRACTIONS

Draw a monster face with 6 eyes and 4 noses. Color 1/2 of the eyes purple. Color 1/4 of the noses orange.

(99)

PLACE VALUE

How many tens are in 100?

(100)

Bonus Builder #20

You cut and use 4 centimeters of this yarn. How much yarn is left?

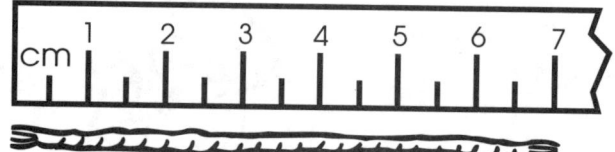

MEASUREMENT

Geometry

How are a cone and a cylinder alike?

101

Problem Solving

How many tricycles are there if you see 15 wheels?

102

Graphing, Probability, & Statistics

If each △ = 1 hour, which 2 people took 5 hours all together to complete their chores?

Time Spent on Chores

Ike △ △ △
Sara △ △
Wendy △ △ △ △

103

Fractions

What fraction of the octagon is white?

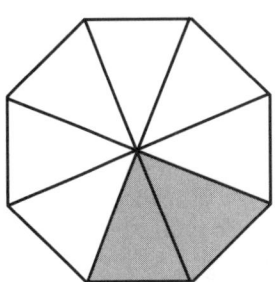

104

Measurement

Which is longer: 1 foot or 11 inches?

105

Bonus Builder #21

Show the number 105 using blocks. How many hundreds, tens, and ones blocks will you need?

Place Value

Time

How many months are there in half a year?

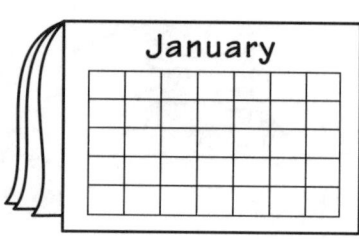

106

Patterns & Relationships

What is the rule for this pattern?

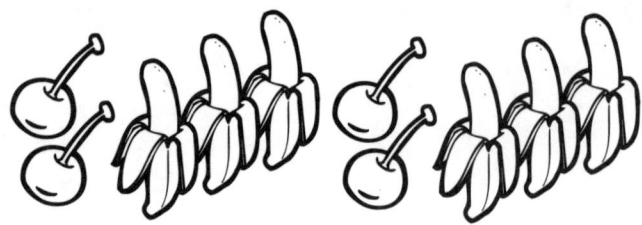

107

Money

How many dimes equal 40¢?

108

Place Value

I have 2 more tens and 4 more ones than the number 35. What number am I?

109

Graphing, Probability, & Statistics

Which number will you probably spin most often in the game? Why?

110

Bonus Builder #22

If you buy 3 comic books each week, how many will you have in 4 weeks?

Computation

Numeration

Draw a picture that shows an odd number of hats.

(111)

Computation

Kip had 16 socks. He lost 2 pairs. How many are left?

(112)

Graphing, Probability, & Statistics

Look at the graph. On which day did Jake earn 5¢ more than on Monday?

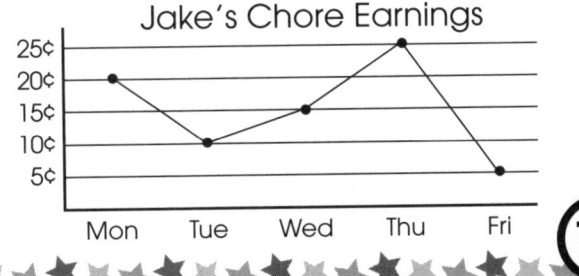

(113)

Problem Solving

Jenna ran the fastest in the race. Todd did not finish last. Who won second place: Jenna, Mary, or Todd?

(114)

Time

How many minutes have passed?

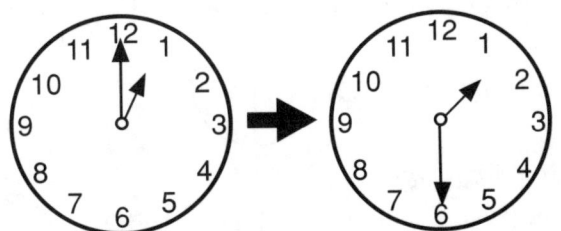

(115)

Bonus Builder #23

Complete the pattern.

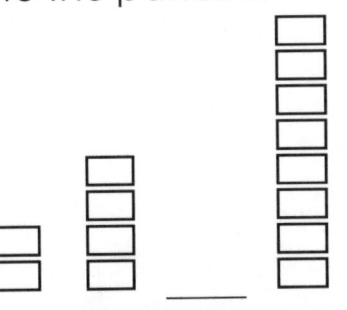

Patterns & Relationships

Money

How much change will you get back?

116

Measurement

How much more milk is needed to make 3 cups?

— 3 cups
— 2 cups
— 1 cup

117

Geometry

Draw 3 things from your classroom that are cube shaped.

118

Fractions

Color 3 beans purple and 2 red. Color the rest yellow. What fraction of the beans are yellow?

119

Place Value

Add 2 tens to the number 16. What is the sum?

120

Bonus Builder #24

Fred and Mike started their homework at .

Fred finished at .

Mike finished in one hour. Who finished first?

Time

Computation

Mom made 17 brownies. José and Victor each ate 3 brownies. How many are left?

(121)

Graphing, Probability, & Statistics

How many students voted for the fish as their favorite pet?

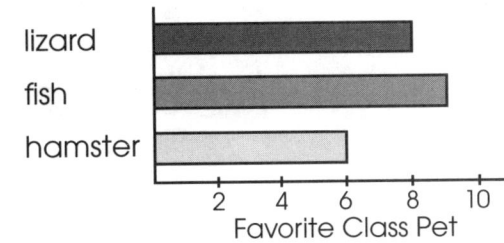

(122)

Numeration

Draw a bug with 6 pairs of feet.

(123)

Patterns & Relationships

What is the rule for this pattern?

(124)

Numeration

Rob's favorite number is less than 12. It is not an even number. It is 2, 9, or 13. Which number is it?

(125)

Bonus Builder #25

How many angles (corners) does an octagon have?

Geometry

Graphing, Probability, & Statistics

With your eyes closed, what is your chance of pulling out the white piece of candy?

____ out of ____

126

Fractions

What fraction of these toys are *not* broken?

127

Money

Which amount is greater: $1.10 or 103¢?

128

Measurement

One glass holds 4 ounces of juice. How many glasses will 12 ounces of juice fill?

129

Problem Solving

Color the path that ends with 10.

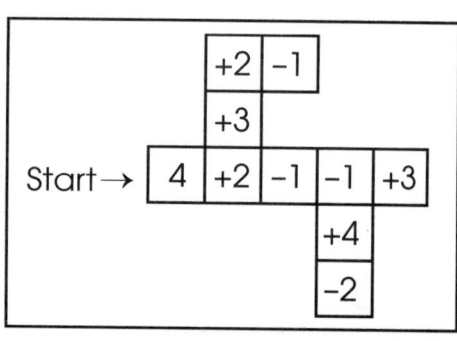

130

Bonus Builder #26

I am an even number between 17 and 24. I have a 0 in the ones place. What number am I?

Numeration

Place Value

What is the largest two-digit number you can make with a 4 in the tens place?

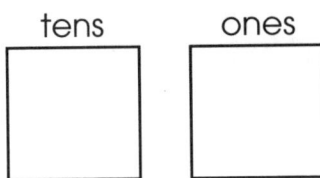

131

Time

If it takes 30 minutes to play a game, how many times can you play it between 1:00 P.M. and 3:00 P.M.?

132

Patterns & Relationships

Complete the following pattern.

△○□△○□ ○□

133

Graphing, Probability, & Statistics

Complete the graph. Show Pal with 3 more bones than Fifi. Show Duke with one less bone than Pal.

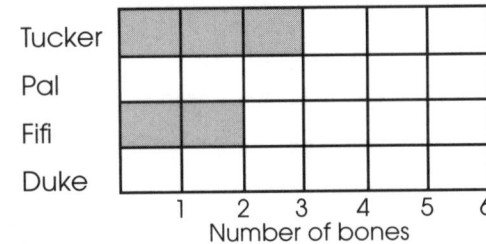

134

Computation

There were 15 home runs hit this season. Danny hit 6 of these, and Dee hit the rest. How many did Dee hit?

135

Bonus Builder #27

How many ladybugs end to end would equal 10 centimeters?

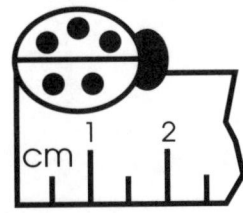

Measurement

Geometry

Which of these are congruent (the same size and shape)?

A. B. C. D. E.

136

Measurement

What is the temperature?

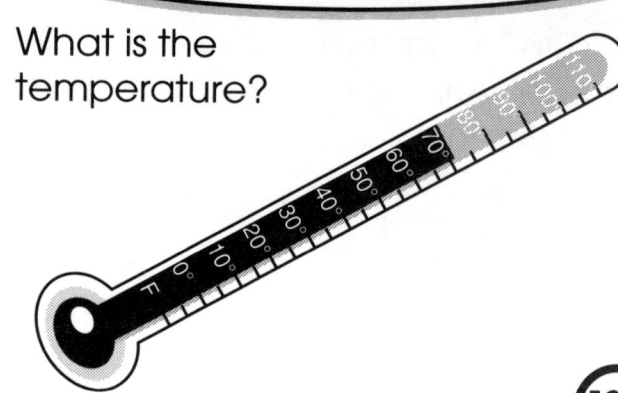

137

Graphing, Probability, & Statistics

If each ◯ = 3 books read, how many more books did Randy read than Ted?

Sue   ◯◯◯◯
Ted   ◯◯
Randy ◯◯◯◯◯◯

138

Problem Solving

How many marbles will go into each box if you want the same number in each?

139

Fractions

What fraction of the cake has been eaten?

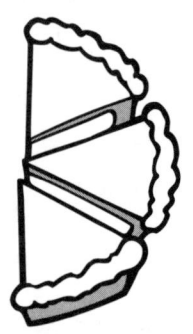

140

Bonus Builder #28

Which 2 sets have the same amount of money?

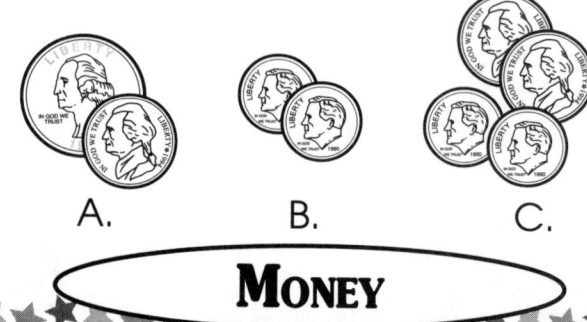

A.   B.   C.

Money

Fractions

Draw a pie cut into 6 equal parts. Color 1/3 of the slices.

(141)

Time

How many months are there in 2 years?

(142)

Numeration

Color the fifth car blue and the third car red.

(143)

Patterns & Relationships

Complete this number pattern.

0, 3, 6, 9, ___, 15

(144)

Geometry

I am a solid shape with 1 flat square side and 4 flat triangle sides. What am I?

(145)

Bonus Builder #29

How many more people like the color blue than purple?

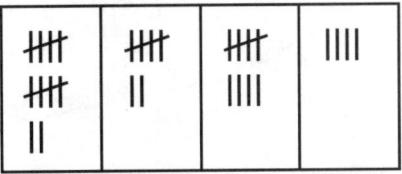

Graphing, Probability, & Statistics

Computation

If you buy 3 more fish and then give 1 away, how many fish will you have?

146

Problem Solving

How many people are there if each person gets a pair of gloves?

147

Money

What is 3¢ less than this?

148

Graphing, Probability, & Statistics

Which dessert is the second favorite?

Favorite Desserts

149

Measurement

Which would weigh more?

150

Bonus Builder #30

How many tens and ones can you make?

Place Value

Money

How much change will you get back?

151

Time

What time was it 2 hours earlier?

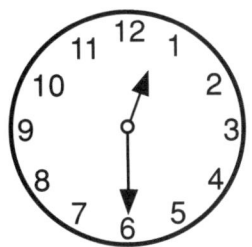

152

Graphing, Probability, & Statistics

If you pick a card without looking, what is your chance of picking a card with an 8?

____ out of ____

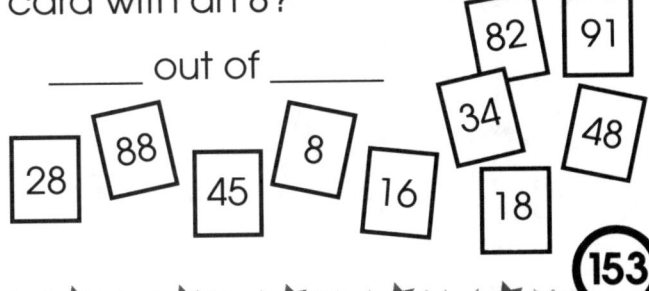

153

Geometry

How many square sides does a cube have?

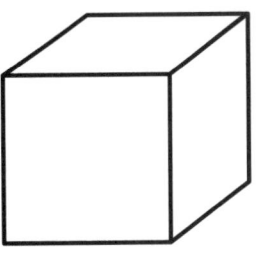

154

Fractions

Four children want to share a cookie. Draw 1 cookie divided equally.

155

Bonus Builder #31

What 2 numbers have a sum of 13 and a difference of 1?

Numeration

Write the first, sixth, and seventh letters of the word *helicopter*. What word did you spell?

(156)

Place Value

Write the number for each set.

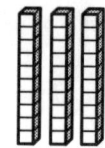 

_____   _____

(157)

Patterns & Relationships

What number continues this pattern?

1, 4, 7, 10, ____

(158)

Problem Solving

There were 14 bananas. The monkey has eaten 2 each day. How many days has he eaten?

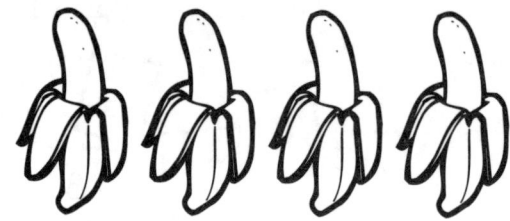

(159)

Measurement

How many inches shorter is the pencil than the scissors?

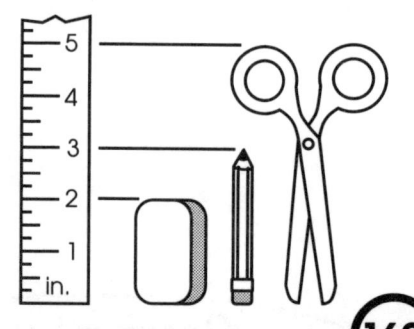

(160)

Bonus Builder #32

Draw a graph using these facts:
- Two kids like ice cream the best.
- Cake is liked the most by 4 kids.
- None like pie best.

Graphing, Probability, & Statistics

Problem Solving

Cindy doubles the number of blocks used each day. How many will she use on the third day?

(161)

Computation

You have 16 marbles, and your friend has 42. How many do you have all together?

(162)

Place Value

I am a number with 2 fewer tens, but 8 more ones. What number am I?

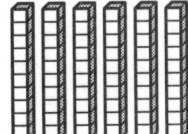

(163)

Money

Which amount is less?

A.    B.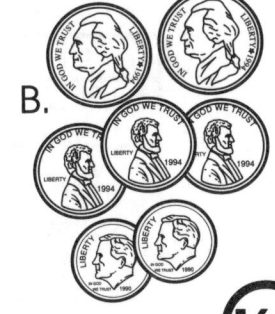

(164)

Patterns & Relationships

What comes next?

June, August, October, _____

(165)

Bonus Builder #33

What time will it be in 1 hour and 30 minutes?

Time

Geometry

I am a number inside the square but outside the circle. I am also inside the triangle. What number am I?

166

Measurement

If there are 4 cups in a quart, how many cups are there in 2 quarts?

167

Time

Today is February 28. What was the date 17 days ago?

FEBRUARY						
S	M	T	W	T	F	S
	1	2	3	4	5	6
7	8	9	10	11	12	13
14	15	16	17	18	19	20
21	22	23	24	25	26	27
28						

168

Graphing, Probability, & Statistics

How many more kids own 🛼 than 🚲?

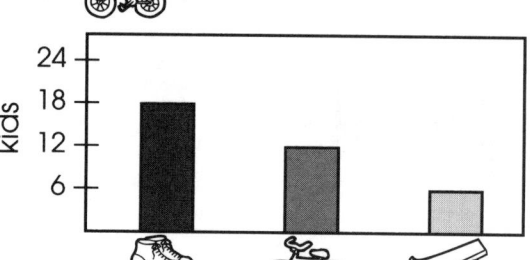

169

Numeration

Write the number that comes before and after each number.

___ 18 ___

___ 26 ___

___ 30 ___

170

Bonus Builder #34

Four kids were playing hockey. Two left. What fraction of the children left the game?

Fractions

Computation

You scored 18 points for your team. Your friend scored 12 points. How many more points did you score?

(171)

Fractions

Draw this sandwich cut into 4 equal pieces. Color 3/4 of the pieces yellow.

(172)

Money

Add 2 dimes to this amount. What is the new total?

(173)

Patterns & Relationships

Continue the pattern. Write the rule.

31, 28, 25, 22, 19, ____, ____

(174)

Graphing, Probability, & Statistics

Which number will you probably stop on least often? Why?

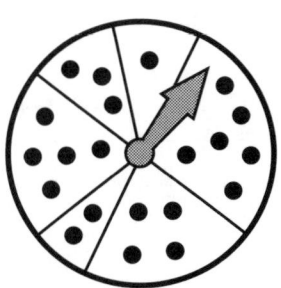

(175)

Bonus Builder #35

One nickel buys 2 pieces of gum. How many pieces can you get for a quarter?

Problem Solving

Geometry

Name 2 solid shapes that have a flat face that is a circle.

(176)

Measurement

How many degrees cooler was it on Sunday than on Friday?

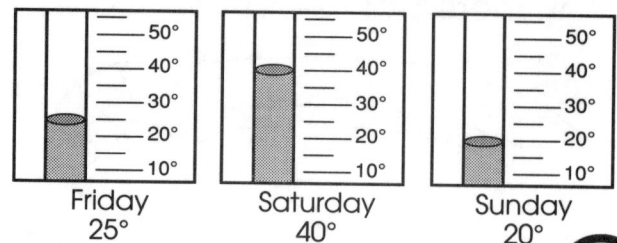

Friday 25°   Saturday 40°   Sunday 20°

(177)

Place Value

How many of each of the following are in the number 178?

_____ hundreds
_____ tens
_____ ones

(178)

Time

How many minutes' difference is there between the 2 clocks?

(179)

Graphing, Probability, & Statistics

How many kids play a sport after school?

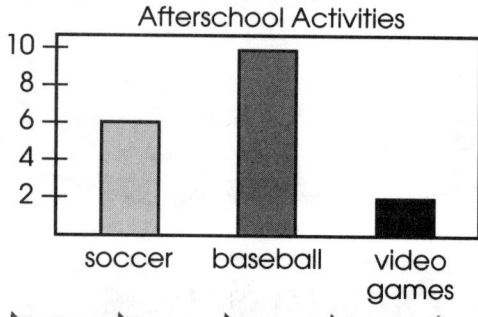

(180)

Bonus Builder #36

Complete this pattern.

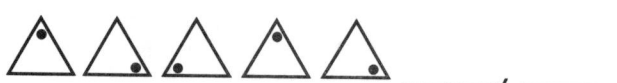

Patterns & Relationships

Graphing, Probability, & Statistics

Write 3 things you know after reading this chart.

Number Rolled	Times Rolled
6	III
4	II
3	IIII I
1	II

181

Time

Write the day and the date 1 day after Wednesday, September 13.

182

Money

Which 3 coins equal 16¢?

183

Fractions

Write a sentence about this picture. Use a fraction.

184

Computation

At the zoo you saw 11 zebras, 5 lions, and 3 kangaroos. How many animals did you see in all?

185

Bonus Builder #37

Name 3 solid shapes that have a flat face that is a square.

Geometry

Place Value

How many tens and ones can you make?

(186)

Fractions

What fraction of the eggs is missing?

(187)

Problem Solving

Twenty kids want to play a game. Each team needs 4 kids. How many teams will there be?

(188)

Measurement

How many inches around is this box?

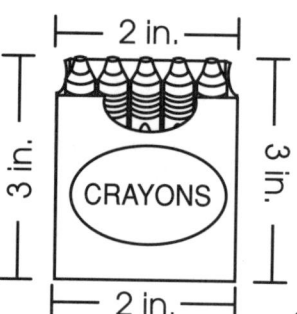

(189)

Computation

If there are 5 bikes and 2 trikes, how many wheels are there in all?

(190)

Bonus Builder #38

What 4 coins are in the bank?

Money

Computation

Write a number sentence to show the total number of frogs.

(191)

Geometry

Which of these is a closed shape?

(192)

Money

What coin would be equal to the amount shown?

(193)

Place Value

What is the biggest number you can make using these 3 numbers?

1   6   4

(194)

Time

Lunch begins at 12:00 and ends in 1 hour. What time does lunch end?

(195)

Bonus Builder #39

Name the fraction for 1 slice of pie.

Fractions

Patterns & Relationships

Use the pattern to draw a necklace that has 9 beads.

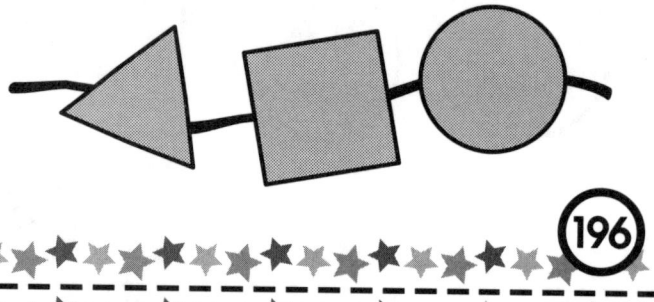

196

Problem Solving

Joe has 10 bugs. He puts the same number in each tank. How many bugs are in each tank?

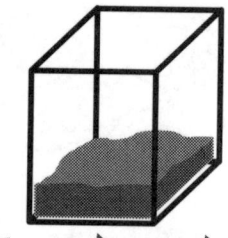 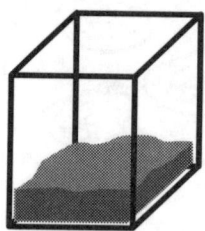

197

Numeration

Put these numbers in order from the least to the greatest.

43   10   27   34

198

Measurement

Would you use inches, pounds, or cups to measure the weight of this fruit?

199

Graphing, Probability, & Statistics

How many more students like apples than pears? How many students like pears and oranges all together?

Fruits	Number of Students
apple	
pear	
orange	

200

Bonus Builder #40

Write 3 more number sentences using the numerals 2, 6, and 8.

1. 2 + 6 = 8
2.
3.
4.

Computation

Geometry

Name a solid shape with no flat sides.

201

Money

Amy bought gum for 10¢. What 2 coins did she give the clerk?

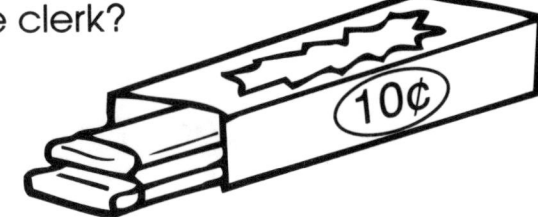

202

Place Value

Write each number.

8 tens and 2 ones
4 tens and 1 one
6 tens and 5 ones

203

Computation

Write a number in each ◯ to make each problem correct.

204

Fractions

What fraction of the cake is missing?

205

Bonus Builder #41

Continue this pattern. Write the rule.

100, 200, 300, ____, ____

Patterns & Relationships

Time

How many minutes long was the TV show?

Start  Finish

206

Computation

Which number sentence is wrong?

$5 + 8 = 13$

$6 + 7 = 12$

$12 - 6 = 6$

207

Numeration

List 5 odd numbers less than 20.

208

Money

List the coin names in order from least to greatest in value.

209

Patterns & Relationships

What would the next number sentence be?

$3 + 1 = 4$

$3 + 2 = 5$

$3 + 3 = 6$

__ + __ = __

210

Bonus Builder #42

Which of these shapes is not divided into fourths? Explain your answer.

 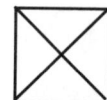 

Fractions

Answer Keys

Page 3
1. The student should have drawn 6 squares.
2. △ □ △ □ △ □
3. 6 birds
4. 8 circles
5. the student

Bonus Builder #1: 10 shells

Page 4
6. 10 cookies
7. 9¢
8. 7 days
9. the spider, 4 more legs
10. 4

Bonus Builder #2: Student answers will vary, but can include that Tom has more bears, Tom has 1 more bear, Sara has fewer bears, etc.

Page 5
11. noon and midnight
12. 6, 7, 8
13. Sunday
14. 11
15. 3 flowers

Bonus Builder #3: 10 paper clips wide

Page 6
16. 3 pairs
17. The student should have circled the dime.
18. eating
19. The student should have drawn a triangle.
20. 5:00

Bonus Builder #4: 13

Page 7
21. a nickel and a penny
22. 5 blocks longer
23. 2
24. a circle
25. 19

Bonus Builder #5: 5:00

Page 8
26. 7 more shoes
27. The student may have drawn □ , △ , or ▱ .
28. 14 days
29. The student should have drawn a large circle.
30. the smiley faces and the apple

Bonus Builder #6: the black jelly beans, because there are more of them

Page 9
31. 57
32. Joe
33. 8¢
34. 8 centimeters
35. 4 slices

Bonus Builder #7: 2 more baskets

Page 10
36. The student should have drawn a small circle.
37. 9 games
38. 9 girls
39. The student should have drawn 3 different ways: 2 nickels, 1 nickel and 5 pennies, and 10 pennies.
40. 10 marbles

Bonus Builder #8: 10, 12, 8, 16

Page 11
41. The student should have drawn 2 butterflies.
42. a ruler
43. Student drawings will vary.
44. 16¢
45. Student drawings will vary, but 4 windows should be colored yellow.

Bonus Builder #9: yes

Page 12
46. 23¢
47. Tina and Jo
48. ⟢
49. 2 tens and 4 ones
50. Student drawings will vary, but 1 of the birds should be colored blue.

Bonus Builder #10: 13¢

Page 13
51. 6 pennies
52. Tuesday
53. 16¢
54. 2
55. Student drawings will vary, but the caterpillar should be 5 centimeters long.

Bonus Builder #11: The student should have drawn a stack of 4 boxes.

Answer Keys

Page 14
56. 15¢
57. the diamond, because the diamond shape has the largest space on the spinner
58. 5, 11
59. 21
60. 6 tens
Bonus Builder #12: $1/4$

Page 15
61. 10:00
62. Student drawings will vary, but girls should outnumber boys by 1.
63. weight
64. 3
65. 6 more cards
Bonus Builder #13: Student drawings will vary and can be either a pictograph, bar graph, or pie graph showing themselves asleep 10 hours and awake 14 hours.

Page 16
66. third place
67. 12¢
68. Student answers will vary.
69. 7 rectangles
70. 6 halves
Bonus Builder #14: 4 cookies

Page 17
71. a basketball, because it is the ball that goes with the identified sports equipment
72. the watermelon
73. 6 kids
74. Students should have drawn 4 nickels.
75. 2
Bonus Builder #15: 1:30

Page 18
76. Student answers will vary, but may include a square, a rectangle, and a diamond.
77. 152
78. no
79. Student answers will vary, but should include 2 of the following: $4/8$, $1/2$, $2/4$.
80. 1, 3, 5, 7, 9
Bonus Builder #16: B

Page 19
81. 2 more blocks
82. 4:30
83. 10 nickels
84. the baseball
85. 8
Bonus Builder #17: 13

Page 20
86. 26, because each of the other numbers has a 1 in the tens place
87. $1/4$
88. 4 peanuts
89. round
90. 8
Bonus Builder #18: 7 sides

Page 21
91. 4, 4
92. Student answers will vary.
93. Student answers will vary.
94. 60 minutes
95. 5 fish
Bonus Builder #19: 18¢

Page 22
96. 22, 24, 26, 28, 30, 32
97. Sue and Cal
98. 3 mother cats
99. Student drawings will vary, but 3 of the eyes should be purple and 1 of the noses should be orange.
100. 10 tens
Bonus Builder #20: 3 centimeters

Answer Keys

Page 23
101. Student answers will vary.
102. 5 tricycles
103. Ike and Sara
104. 6/8 or 3/4
105. 1 foot
Bonus Builder #21: 1, 0, 5

Page 24
106. 6 months
107. The pattern alternates between 2 cherries and 3 bananas.
108. 4 dimes
109. 59
110. 5, because it has the largest space on the spinner
Bonus Builder #22: 12 comic books

Page 25
111. Student drawings will vary, but should show an odd number of hats.
112. 12 socks (or 6 pairs)
113. Thursday
114. Todd
115. 30 minutes
Bonus Builder #23: The student should have drawn a stack of 6 blocks.

Page 26
116. 10¢
117. 1 cup
118. Student drawings will vary.
119. 5/10 or 1/2
120. 36
Bonus Builder #24: Fred

Page 27
121. 11 brownies
122. 9 students
123. Student drawings will vary, but should include a bug with 12 feet.
124. One more spot is added to the ladybug's wings each time.
125. 9
Bonus Builder #25: 8 angles (corners)

Page 28
126. 1 out of 8
127. 3/4
128. $1.10
129. 3 glasses
130.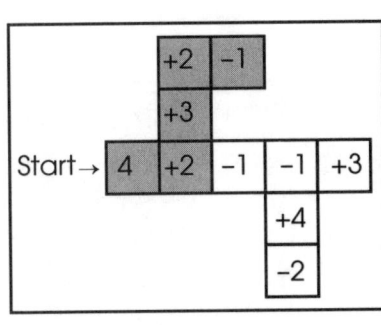
Bonus Builder #26: 20

Page 29
131. 49
132. 4 times
133. The student should have drawn a triangle.
134.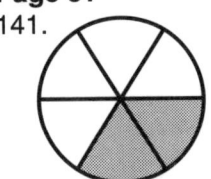
135. 9 home runs
Bonus Builder #27: 5 ladybugs

Page 30
136. A and D
137. 71°F
138. 12 more books
139. 5 marbles
140. 1/2 or 3/6
Bonus Builder #28: Sets A and C

Page 31
141.

142. 24 months
143. The student should have colored the third car red and the fifth car blue.
144. 12
145. a pyramid
Bonus Builder #29: 3 more people

Page 32
146. 12 fish
147. 5 people
148. 22¢
149. cake
150. 4 quarters
Bonus Builder #30: 7 tens and 1 one

Page 33
151. 5¢
152. 10:30
153. 6 out of 10
154. 6 square sides
155. Student drawings will vary, but should show a cookie cut into 4 equal pieces.
Bonus Builder #31: 7 and 6

Answer Keys

Page 34
156. *h, o, p; hop*
157. 33, 49
158. 13
159. 5 days
160. 2 inches
Bonus Builder #32: Student drawings will vary and can be either bar graphs, circle graphs, or pictographs. They should show that 2 kids prefer ice cream, 4 prefer cake, and none prefer pie.

Page 35
161. 8 blocks
162. 58 marbles
163. 49 (4 tens and 9 ones)
164. Set B
165. December
Bonus Builder #33: 4:30

Page 36
166. 6
167. 8 cups
168. February 11
169. 6 kids
170. 17, 19; 25, 27; 29, 31
Bonus Builder #34: $2/4$ or $1/2$

Page 37
171. 6 more points
172. The student should have shown the sandwich cut into 4 pieces with 3 of them colored yellow.
173. 64¢
174. 16, 13; the pattern counts backward by 3.
175. 2, because it has the smallest space on the spinner
Bonus Builder #35: 10 pieces

Page 38
176. a cone and a cylinder
177. 5°F
178. 1 hundred, 7 tens, and 8 ones
179. 30 minutes
180. 16 kids
Bonus Builder #36: △ △

Page 39
181. Student answers will vary, but may include that the numbers 4 and 1 were rolled the same number of times, the number 3 was rolled the most times, etc.
182. Thursday, September 14
183. a dime, a nickel, and a penny
184. Student sentences will vary, but should include $2/5$ or $3/5$.
185. 19 animals
Bonus Builder #37: a cube, a pyramid, and a box (rectangular prism)

Page 40
186. 7 tens and 2 ones
187. $4/12$ or $1/3$
188. 5 teams
189. 10 inches
190. 16 wheels
Bonus Builder #38: 2 dimes and 2 nickels

Page 41
191. $3 + 5 = 8$
192. the circle
193. a quarter
194. 641
195. 1:00
Bonus Builder #39: $1/4$

Page 42
196. -▲-■-●-▲-■-●-▲-■-●-
197. 5 bugs
198. 10, 27, 34, 43
199. pounds
200. 4 more students, 9 students
Bonus Builder #40: $6 + 2 = 8$, $8 - 6 = 2$, $8 - 2 = 6$

Page 43
201. a sphere
202. 2 nickels
203. 82, 41, 65
204. 2, 4, 4
205. $1/4$
Bonus Builder #41: 400, 500; 100 is added each time.

Page 44
206. 30 minutes
207. $6 + 7 = 12$
208. Answers will vary, but should include 5 of the following: 1, 3, 5, 7, 9, 11, 13, 15, 17, 19.
209. penny, nickel, dime, quarter
210. $3 + 4 = 7$
Bonus Builder #42: the triangle, because it is not equally divided